An A to Z of Inventions and Innovations
that Changed the World

Denise Morgan Branch

Dedication

This book is dedicated to my children, Royce, Brandon and Ashley with love for their support throughout the writing of this book, and in memory of my parents Promise and Ethel Morgan for encouraging my natural curiosity for books and a love of learning.

Acknowledgments

A very special thanks to Antonia Prescott for providing outstanding consulting and editing services and believing in me throughout this project.

Thanks to Shay Page, my freelance illustrator, for doing a creative job on the illustrations and cover design.

Introduction

Have you ever thought about who the inventors or innovators were for any of the products that you use every day? What was their motivation to create the inventions or innovations? What problem did they set out to solve? Can you imagine a world without these inventions or innovations? The book you hold in your hands captures a glimpse of some of the inventions and innovations that have changed our lives throughout the ages. Readers will step back in time and discover how inventors and innovators pursued an idea and made it a reality that forever changed our world. Without their tenacity, many of the inventions and innovations probably would not exist today.

As a parent, educator, and former homeschool mother, it was my goal to write a book for parents and educators to help them motivate and cultivate a desire for learning how science, technology, engineering, and mathematics led to the creation of inventions and innovations that have an impact on our lives daily. Reading about these inventions and innovations will help students to understand how important these subjects are to problem solving. There are many problems that need to be solved today such as shortages of food and clean water, healthcare, education, climate change, urban planning, energy and many more. By learning how these inventors and innovators solved problems, the reader can begin to think about ways to solve the problems we face today.

An A to Z of Inventions and Innovations that Changed the World was written to expose young minds to how problems of the past were solved by creating new inventions or innovations and to stir their curiosity through critical thinking on how to solve future problems. Since there is so much talk today in education about science, technology, engineering, and mathematics (STEM), students need to develop an interest in and a passion for these subjects, in a fun and exciting way, early in their school career.

The reader will learn how the bulldozer was created from junkyard parts, why the yield sign was developed for a high-traffic area, and how an inventor disobeyed the instructions of his boss and created the UPC (Universal Product Code). Since the book is interactive, at the end of each brief history, the reader is asked to answer carefully selected critical thinking questions about each invention or innovation. Each set of questions challenges the reader to think about each invention or innovation.

Are you ready? Let's take a journey through the pages of history and learn how science, technology, engineering, and mathematics helped inventors and innovators solve problems and forever changed our lives!

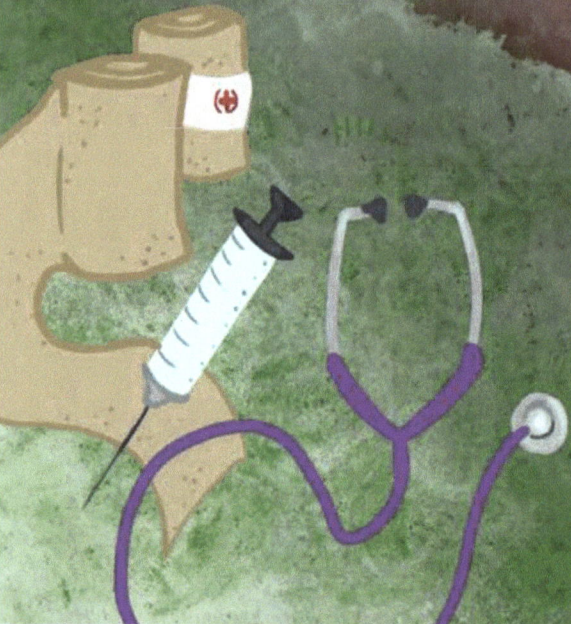

In 1947, the first air ambulance launched to service in the Unit[ed] [W]alter Schaefer was an innova[tor] [ambu]lance tran[sport]

Ambulance

As far back as the eleventh and twelfth centuries, ambulances were used to transport injured soldiers. In those days, the ambulances were horse-drawn wagons. During the Crusades, these wagons moved soldiers from the battlefield to hospital tents.

Walter Schaefer was an innovator in ground and air ambulance transport. He founded Schaefer Ambulance Service in Los Angeles, California in 1932. In 1947, he launched the first air ambulance service in the United States. Schaefer Ambulance Service was the first to provide critical care transportation for patients.

Headquartered in Los Angeles, California, the service was family owned and operated. The company provided excellent medical care for patients with its team of highly qualified staff and sophisticated equipment. After over eight decades of service, Schaefer Ambulance closed in 2019.

Questions

Do you think the ride in the ambulances in the eleventh or twelfth centuries was comfortable for a wounded soldier? Why? Have you ever seen the inside of an ambulance? Describe what you saw. Can you think of anything that should be added?

Bulldozer

James Cummings, an American farmer, invented the bulldozer when the "oil highway" or pipeline was being laid from Wyoming to Missouri in the 1920s by the Sinclair Oil Company. Digging machines were used to hollow out the trenches for the pipeline, but refilling the trenches with dirt was still being done using mules and dirt slings.

Cummings knew there had to be a more efficient way to refill the trenches with dirt after the pipes had been laid. He and J. Earl McLeod, a draftsman, designed the plans for a bulldozer and built the first one from junkyard parts in 1923. They used the frame of a Model-T Ford, some old windmill springs and other junk parts from cars and tractors. Their bulldozer won them a contract with the Sinclair Oil Company to backfill the trenches to the end of the pipeline in Freeman, Missouri.

Questions

Can you imagine what it was like to build a new invention from junkyard parts? Have you ever built anything from junk parts? What did you use? Did your invention work? What problems did you encounter?

KANSAS MORROWVILE
NEWS

1923

James Cummings and J. Earl McLeod, a draftsman, designed the plans for a bulldozer and built the first e from junkyard parts in 1923. They used the frame Model-T, some old windmill springs, and other parts from cars and tractors. Their bulldozer won contract with the Sinclair Oil Company to e trenches to the end of the pipeline in

Cable Television

In 1947, John and Margaret Walson were the owners of Community Antenna Television in Mahanoy City, Pennsylvania. They were having trouble selling televisions, because the eastern Pennsylvania Mountains interfered with reception. So, John Walson installed an antenna on the top of the New Boston Mountain and built a tower. This provided better television reception, and television sales increased.

In 1948, Walson connected customers' homes along the cable path. He charged $100 for installation and $2 a month for the service. This novel approach marked the beginnings of cable television.

Questions

Do you think cable TV would have been invented if Walson's customers had already had good television reception? Why? Compare the price of Walson's cable television installation and monthly service charge to the cost today. Has it increased? Why do you think that might be?

Dishwasher

Josephine Cochran, an Illinois socialite, enjoyed entertaining. However, the servants often chipped her fine china while hand washing it after her social engagements. Josephine was tired of this, so she decided to invent a machine to wash the dishes instead. Her machine consisted of wire compartments for plates, cups, and saucers. These compartments were placed inside of a wheel that lay flat in a copper boiler. The wheel was turned by a motor and hot soapy water was pumped out on the dishes.

Josephine received a patent for her dishwasher on December 28, 1886. She established one of the first kitchen appliance companies in America. Today's KitchenAid dishwasher is a descendant of Josephine's early dishwashing machines. Thanks to Josephine's determination, many households have dishwashers today.

Questions

Why was Josephine so upset when the servants chipped her china? Can you imagine what it would be like today without dishwashers? Have you ever broken one of your parents' dishes? How did they feel? How did you feel?

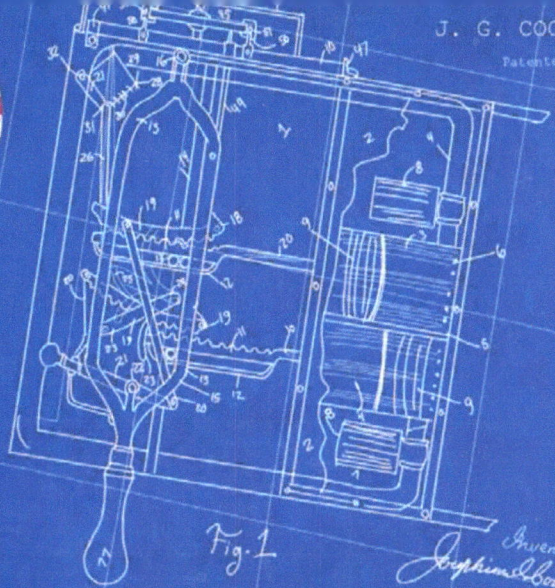

J. G. COCHRAN
Patented Dec. 28, 1886

Fig.1

Inventor:
Josephine G. Cochrane

Escalator

During the early 1890s, Jesse Reno and Charles Seeberger were working separately on their own design for an escalator. Reno's plan was for a slanted conveyor belt that would rise at a 25-degree angle. People would step on to the belt from the floor and be carried to the next level. Seeberger's design involved a flat step with a side entrance, and was a moving staircase. However, neither design had moving handrails.

The Otis Elevator Company obtained the rights to the designs of Seeberger and Reno in the 1890s, but the company did not combine the plans until the 1920s. This combination became the design for the modern-day escalator.

In 1899, the Otis Elevator Company installed several escalators in their factories for company use. They installed the first public escalators for the New York City elevated railroad and exhibited the invention at the 1900 Paris Exposition, where it won first prize. Some of their other early installations of escalators included department stores in Chicago, Illinois and Philadelphia, Pennsylvania. The Otis Elevator Company became a world leader in the manufacturing of escalators.

Questions

Do you think that if Reno and Seeberger had worked together the escalator would have been invented sooner? Why? How is the escalator similar to or different from the elevator? What might have happened if the escalator had not been invented?

Ferris Wheel

George Washington Gale Ferris Jr., a civil engineer, designed the first Ferris Wheel for the Chicago World's Columbian Exposition in 1893. In 1892, Daniel Burnham, the planner for the Exposition, sent out a request for a new design to rival the Eiffel Tower, which had appeared in the Paris Exposition three years earlier. Ferris, responding to Burnham's request, designed a giant upright wheel. It is said that Ferris created the design on a napkin while having dinner at an engineer's banquet. His design consisted of a 250-foot wheel with thirty-six wooden cars that could seat sixty people each. This allowed 2,160 people to take a ride and view the whole fair at any one time. It must have been quite an exciting ride!

As of today, the world's tallest Ferris wheel is the High Roller located in Las Vegas, Nevada, and it opened to the public in 2014. It stands 550 feet tall, is 520 feet in diameter, and has twenty-eight glass pods that can seat up to forty people each.

Questions

Have you ever seen a Ferris wheel? Have you ridden on one? If so, was it fun? Who rode with you? How many people were in the car? Can you design a new amusement park ride?

250ft tall

36 cars - 60 seating per car

RRIS WHEEL SOUVENIR.

WORLD'S FAIR 1893

WALTZ-POLN

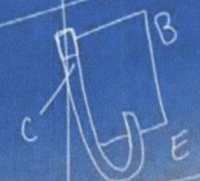

Gas Mask

Garrett Augustus Morgan, an African-American, invented the gas mask in 1912. It was called the "safety hood" or the "breathing device." Morgan had witnessed firefighters struggling to breathe after they entered smoke-filled buildings, so he designed a canvas hood that could be placed over the head. It had two tubes: one supplied fresh air and another discharged the exhaled air. The fresh air tube was coated with a material that could be dampened with water to keep out smoke and dust particles.

Morgan was able to test his device when a terrible explosion occurred at the Cleveland Waterworks in 1916. Workers were trapped in a tunnel under Lake Erie, and previous rescue attempts had failed. Rescuers could not enter the tunnel, because it was filled with poisonous gasses and heavy smoke. Morgan and three volunteers entered the tunnel wearing his gas masks and saved the lives of twenty men.

Questions

What might have happened if the gas mask had not been invented? Can you think of another invention that could help firefighters? How would you design this invention? What special things would you include?

Heart-Lung Machine

John H. Gibbon, Jr., an American surgeon, was inspired to create the heart-lung machine in 1931 after witnessing a woman die from blocked lung circulation. He believed that he could build a machine that could perform the functions of the heart and lungs during heart surgery. This external machine would be able to take deoxygenated blood, oxygenate it, and pump it back into the arterial system.

Gibbon first experimented on animals while working at the University of Pennsylvania School of Medicine in 1939. Later, he became the head of the surgical department at his alma mater, Princeton University. With the backing of Thomas J. Watson, chairman of IBM, he was able to perfect his heart-lung machine. His first operation using the device on a human was unsuccessful, but the second surgery performed on May 6, 1953 worked. This was the first time in history that cardiac surgery was successfully performed on a human.

Questions

What do you think would have happened if the heart-lung machine had not been invented? Did it solve a problem? How? Can you think of a disease or sickness that you would like to cure someday? What device might you need?

1. YGEN-POOR BLOOD LEAVES
HE HEART TO ENTER THE
HEART-LUNG MACHINE

3. OXYGEN-RICH BLOOD RETURNS
TO THE BODY, SKIPPING THE
HEART AND LUNGS

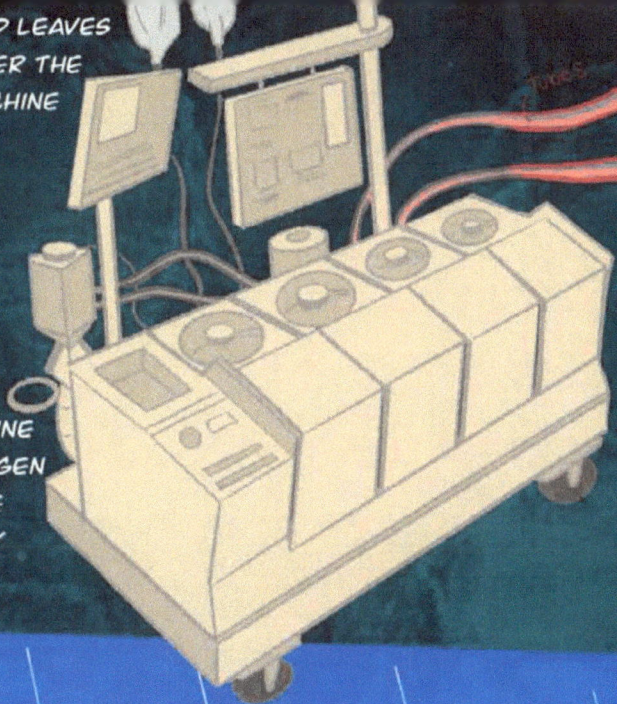

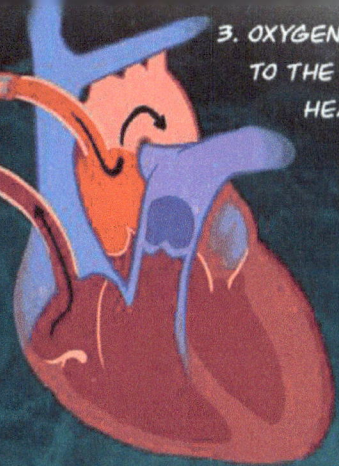

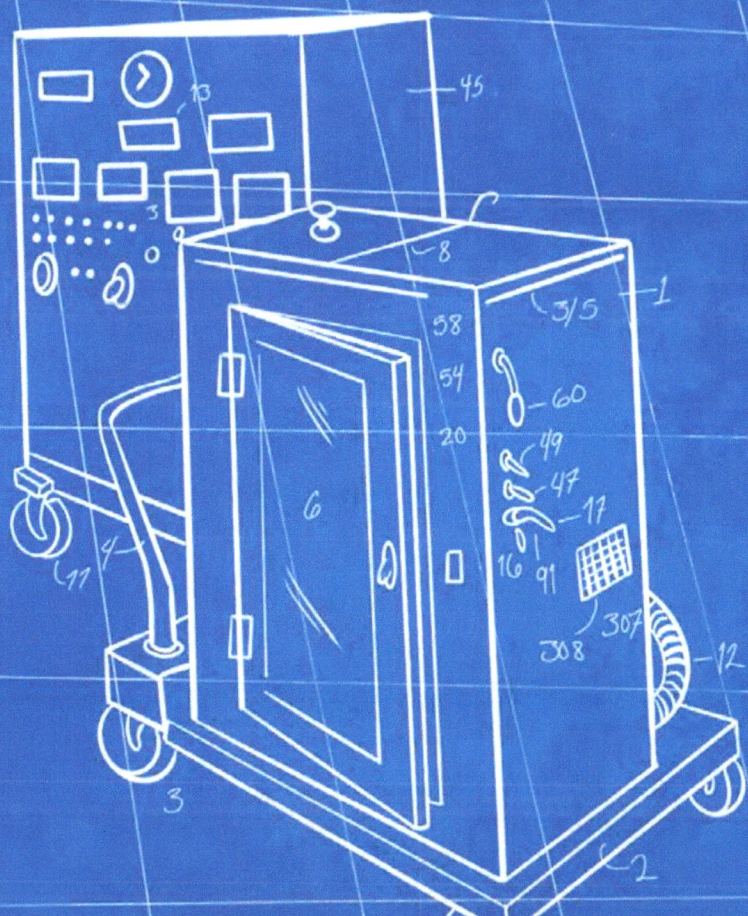

EART-LUNG MACHINE
PS AND ADDS OXYGEN
HE BLOOD BEFORE
URNS TO THE BODY

C.F. KETTERING

Patented Aug, 17, 1915

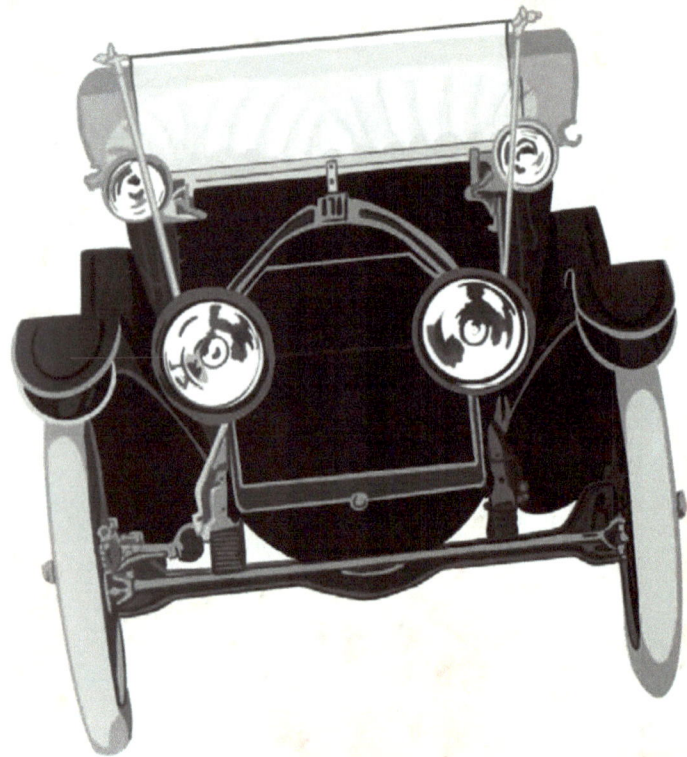

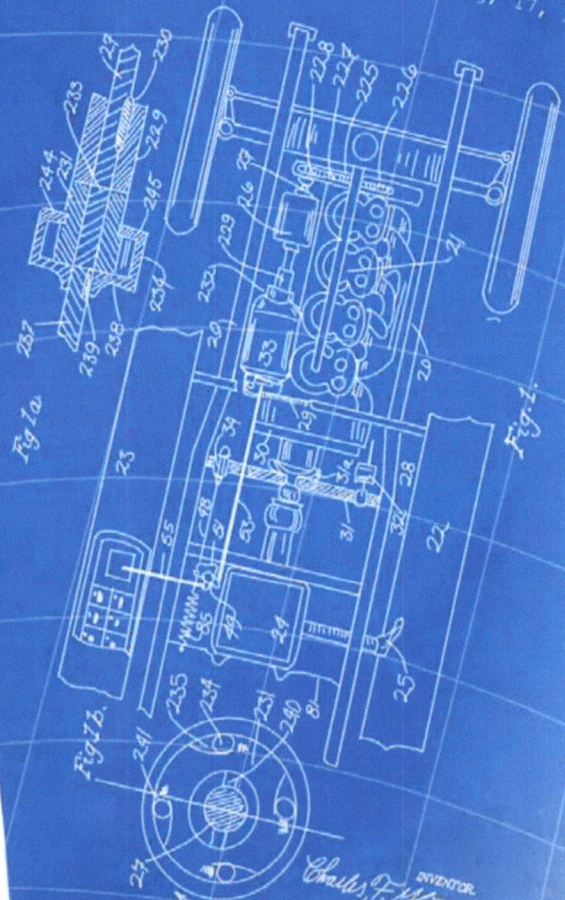

The CAR
THAT HAS NO CRANK

Fig 1a.   Fig 1.

Fig 1b.

INVENTOR.
Charles F. Kettering

Ignition

During the early part of the twentieth century, Charles F. Kettering, an American engineer, invented the electrical ignition system. Prior to this, an iron hand crank had been used to start a car. This was dangerous, because cars did not have safety features on them such as brakes. It was quite common for a car to roll forward or for the engine to backfire.

The first engine starting device was installed in a Cadillac on February 17, 1911. The Cadillac was put up for sale the following year. Kettering received a patent for the ignition system on August 17, 1915. By the 1920s, the electrical ignition system became standard on new automobiles.

Keep on going, and chances are you will stumble on something, perhaps when you are least expecting it. I have never heard of anyone stumbling on something sitting down.

—Charles F. Kettering

Questions

Why was it dangerous to start a car before the invention of the electrical ignition? Can you imagine what it would be like today if cars were still started with a hand crank? Can you think of a new invention for cars today?

Jeans

During the California Gold rush of 1849, miners complained that their pants were wearing out too soon. Long hours of working in the mines caused their trousers to rip easily. In 1850, a dry goods businessman named Levi Strauss moved to San Francisco. He was already making pants out of a cotton fabric called *serge de Nimes*, a French phrase that was Americanized to "denim." Jacob Davis, a tailor, had developed a new method of making trousers out of tent material, using rivets to reinforce the weak spots. However, he could not afford to pay for a patent, so he became partners with Strauss. Together, they received a patent for the new style of jeans on May 20, 1873. Later, jeans became known as "Levi's", so they registered a trademark for the new name.

In addition to being durable work clothes, jeans became fashionable during the 1960s. They have constantly grown in popularity in the fashion industry. By 2007, companies in the United States alone were selling more than 450 million pairs of jeans each year.

Questions

Are the jeans of today designed for durability or fashion? Why? If you were a designer, what changes would you make to the design of jeans? Why? How many pairs of jeans do you think will be sold in the world today?

LEVI STRAUSS & CO.

SPRING BOTTOM PANTS (Riveted)
CAN BE HAD IN EITHER BLUE OR GRAY.

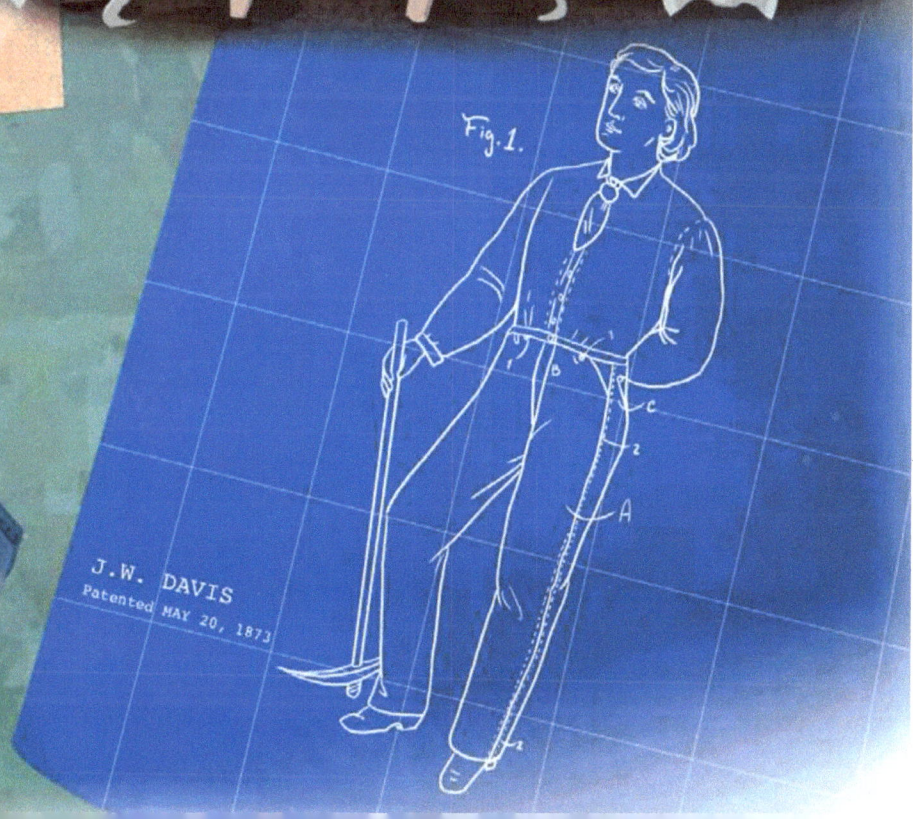

Fig.1.

J.W. DAVIS
Patented MAY 20, 1873

TALKING PICTURES

A FACT!

A REALIT[Y]

THOS. A. EDISON
STARTLES THE CIVILIZED WORLD AND REVOLUTIONIZED T[HE]
PICTURE BUSINESS WITH HIS LATEST AND GREATEST INVEN[TION]
THE KINETOPHONE
ABSOLUTELY THE FIRST PRACTICAL TALKING-MOTION PICTURE EVER
VOICE AND ACTION RECORDED SIMULTANEOUSLY
PERFECT SYNCHRONISMAND ILLUSION
ANY FIRST-CLASS OPERATOR CAN HAND[LE]

THE AMERICAN TALKING PICTURE CO, Inc.

Kinetophone

The turn of the twentieth century was a progressive time in the motion picture industry. Studios in America and Europe were trying to bring together moving images and sound in their motion pictures. In 1895, Thomas Alva Edison successfully combined his Kinetophone (phonograph) with his Kinetoscope (motion picture viewer) and managed to synchronize images with sound. These "talking" movies were played in a cabinet, and viewers could watch motion pictures with sound instead of silent movies. However, the Kinetophone's volume was low, and it only played for about six minutes. Edison hired inventor Daniel Higham in 1908 to help him correct the flaws in the Kinetophone. By 1910, Edison had perfected his device and introduced it to the press.

On February 13, 1913, Edison displayed his Kinetophone during a screening of Shakespeare's *Julius Caesar* in New York. He projected a scene on to a screen, and for seven minutes of the hour-long movie, the audience watched moving pictures with synchronized sound. Edison never fully developed the Kinetophone to produce sound throughout an entire movie. Talking movies were introduced in 1927.

Questions

What do you think would have happened if Edison had not created the Kinetophone? How did it revolutionize the motion picture industry? What is unique about movies today?

Liquid Paper

Bette Nesmith Graham invented liquid paper while working as a secretary at a bank in Texas in 1951. This was a difficult time for women to find a job, especially in the banking industry, but Bette worked her way up to the position of executive secretary.

Bette was not a very good typist, and she used an electric typewriter with carbon film ribbons that made it difficult to erase mistakes. Drawing on her experience as an artist, Bette began using white tempera paint to correct her mistakes. Soon, other secretaries began requesting bottles of Bette's white out. She called it "Mistake Out" and used it for five years before she began to market it in 1956. Bette renamed her product "Liquid Paper" and received a patent for it in 1957. By 1968, Liquid Paper had become a very successful product. Over 10,000 bottles were being produced a day with sales of one million dollars. Liquid Paper steadily increased in production and sales, and Bette sold her company to the Gillette Corporation in 1979 for $47.5 million dollars.

Questions

Why was the invention of liquid paper important for secretaries? How did Bette use her creative ability to improve her job? What might have happened if Bette had not invented liquid paper?

LIQUID PAPER
CORRECTION FLUID

easy t... ...s but hard
...them. In addition, the ribbons
...e with a carbon film which mad...
...ult to erase the mistakes.

...takes... and other secretarie...
...ng bottles of Bette's ...rket it
...it "Mist... and...nt f...
...ears befor... an t. ha...
...1956. ...er t...
...957. By...

Bette Nes...
invented ...
while worki...
secretary a...
in Texas ...

not...

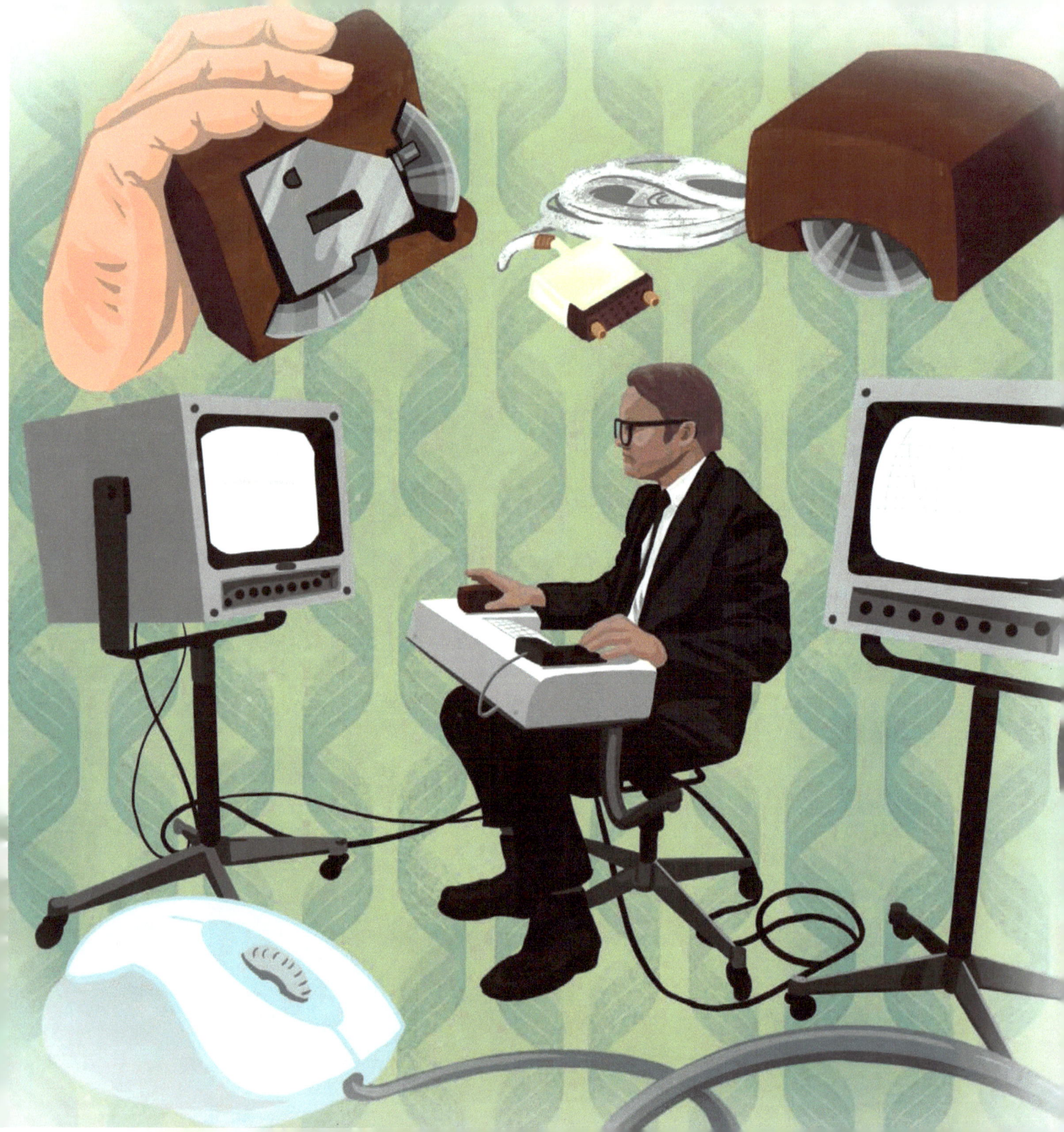

Mouse, Computer

Douglas Engelbart, a computer scientist, invented the first computer mouse during the early 1960s. Early computers were complicated because they had no screens, keyboards, or mouse devices. Engelbart wanted to make computers easy for ordinary people to use. While working as a radar technician, he observed that people looked at things, pointed to them, and moved them with their hands. His original mouse was composed of a wooden case with two wheels inside that rolled along at right angles, attached to the computer by a cable. It was called a mouse because the cable hanging from it looked like a tail.

In 1968, Engelbart gave a 90-minute demonstration in front of an audience of one thousand scientists in San Francisco, California. It was called "The Mother of All Demos." This was an important time in the field of technology, because a year later the internet was launched. He received a patent for the mouse in 1970. Today, the mouse is a standard accessory for all computers.

Questions

What is different about the mouse today compared to the first mouse invented by Engelbart? How has the design of the mouse changed? Why did the mouse make it easier for people to use a computer?

Nylons

In 1928, Dr. Wallace Hume Carothers left his position at Harvard University to join the research department of the DuPont Corporation. The market for silk stockings was worth 70 million dollars in the 1920s. However, silk was becoming harder for the United States to get from Japan because of troubled relations. Carothers' research involved the development of a synthetic material that would be less expensive and could meet the growing demand of the market. Carothers' research led to the invention of nylon, which was patented in 1937.

The first nylon product sold to the public was a toothbrush in 1938. The first nylon stockings were sold in the United States a year later. By 1941, a total of 60 million pairs of stockings were sold in the United States. Nylon is a very strong material, and it is used in clothing, laces, toothbrushes, sails, fish nets, carpets, strings on musical instruments, and many other products.

Questions

Can you think of anything else that is made from nylon? What nylon items do you have at home? Since nylon is a very strong material, can you think of something to invent using it?

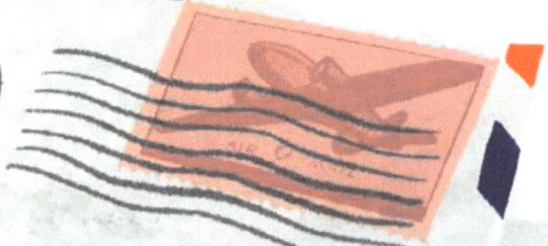

Odometer

Benjamin Franklin, one of the Founding Fathers, took up a position as postmaster of Philadelphia in 1737. Later, in 1775, serving as a member of the Second Continental Congress, he was appointed Postmaster General. Franklin was responsible for developing the postal system in the colonies.

Benjamin Franklin is credited with creating the simple odometer. This machine was used to measure the quickest and most efficient routes between locations for delivering mail. While taking a five-month tour to inspect post offices, Franklin attached a device to the wheel of his carriage, and the distance traveled could be calculated by counting the number of rotations of the wheels and converting these into miles. Today, odometers are used to measure the distances traveled by vehicles, motorcycles, and bicycles.

Questions

Have you ever seen an odometer? Why was the odometer important for the delivery of mail? Do you think the mail delivery time has increased or decreased because of the invention of the odometer? Why?

Potato Chips

George Crum, son of a Native-American mother and an African-American father, was a renowned chef. He created the potato chip on August 24, 1853, while working in a restaurant. A difficult customer returned his fried potatoes several times, complaining that they were too thick and soft. Crum was very upset, so he deliberately sliced the potatoes thinner and fried them harder. To his surprise, the new way of frying potatoes was very popular with the customer. It was so popular that other diners started requesting them too. The next day, the dish appeared on the menu as "George Crum's Saratoga Chips."

The first potato chip production company was opened in Albany, New York, in 1925 by A. A. Walter. Today many kids and adults enjoy eating potato chips as a snack. Even though snacks are fun to eat, they should be eaten in moderation. *Always eat healthy, kids!*

Questions

Why should potato chips be eaten in moderation? Can you name some healthy snacks? Can you think of a new fun healthy snack?

Clean small
wounds

Clean and
dust hard to
reach spaces

Apply cosmetics

Q-TIPS

Q-tips

Leo Gerstenzang, a Polish-born American, invented the Q-tip in 1923. He often watched his wife clean their baby daughter's ears using a tooth-pick with a piece of cotton on the end. This inspired Leo to attach cotton to both ends of a wooden stick. He called his new invention Baby Gays. In 1926, he changed the original name of his invention to Q-tips. In 1948, there was such huge consumer demand for Q-tips that the manufacturing company moved from New York City to Long Island, New York.

Doctors advise never using Q-tips to clean the inside of your ears. They are used for many other purposes such as cleaning and painting, and are commonly found in hospitals, doctors' offices, first-aid kits, and medicine cabinets.

Questions

Why do you think doctors advise not using Q-tips to clean the inside of your ears? Do you think the Q-tip would have been invented if Gerstenzang had not seen his wife cleaning their baby's ears? Why?

Reaper

Cyrus McCormick, a Virginian, invented the reaper in 1831. Prior to this innovation, farm workers could only harvest two to three acres of grain per day. With the advent of the reaper, it was possible to harvest ten acres of grain per day. McCormick continued to improve the technology of the reaper, and by the time he applied for a patent in 1834, it could harvest twenty acres of grain per day, a huge increase on hand reaping.

A threshing device was added to the reaper in the 1800s, which allowed grains to be separated from stalks during harvesting. McCormick made further improvements to the reaper during the 1850s by adding a binder to bundle the grain into bales. The reaper-thresher became the combine. Today a combine can harvest an average of 130 acres per day.

Questions

Have you ever been to a farm? Have you ever seen a modern reaper, also known as a combine? Can you describe it? What is unique about it? Can you think of something new that could be added to the modern-day reaper?

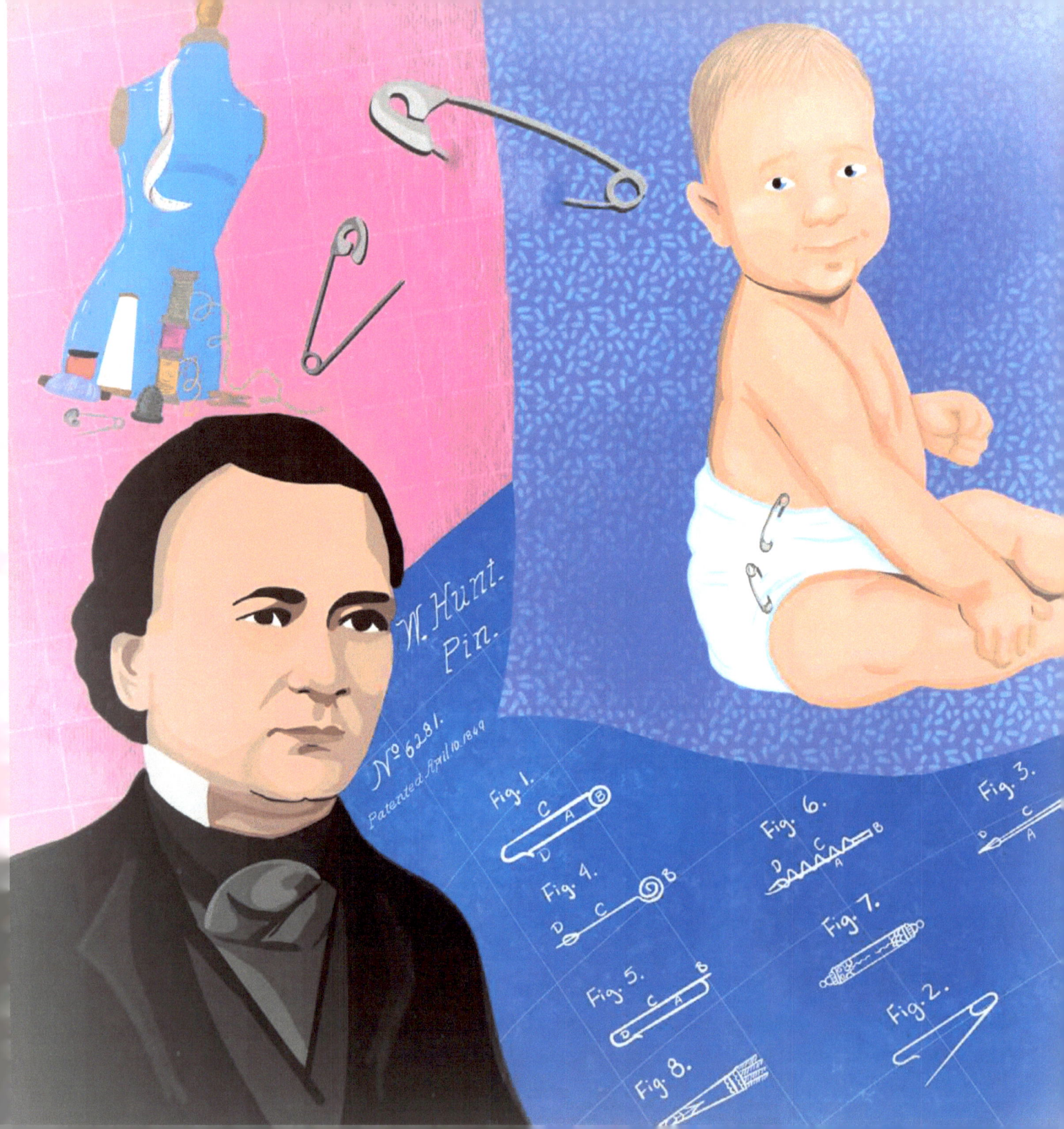

W. Hunt.
Pin.

Nº 6281.
Patented April 10. 1849.

Fig. 1.

Fig. 4.

Fig. 5.

Fig. 8.

Fig. 6.

Fig. 7.

Fig. 2.

Fig. 3.

Safety Pin

The modern-day safety pin was created by Walter Hunt, an American inventor, in 1849. Hunt was twisting a piece of wire while trying to figure out how he was going to pay a $15 debt that he owed to a creditor. While playing with the wire, he was inspired with the idea of making pins safer and preventing people from pricking their fingers. Hunt sold the rights to his invention for $400 so that he could pay off the $15 debt. Hunt's safety pin was patented in April of 1849. Although he was responsible for many other inventions, he sold the rights, so he never profited from their success.

The safety pin has many uses. For example, prior to the invention of disposable diapers, safety pins were used to fasten cloth diapers on a baby. Nowadays, they are found in sewing kits, and are used by tailors and seamstresses.

Questions

Can you think of other ways to use a safety pin? Could the safety pin be improved? Can you design a new safety pin?

Teddy Bear

While on a trip to settle a border dispute between Louisiana and Mississippi in 1902, President Theodore Roosevelt went hunting and spared the life of a bear cub. Clifford Berryman, a political cartoonist, heard what had happened and made a cartoon drawing of the event calling it "Drawing the line in Mississippi."

The event proved to be the inspiration for the teddy bear. When a toy manufacturer called Morris Michtom heard of the president's heroic act, he decided to ask President Roosevelt if he could use the name Teddy for his new toy bear. President Roosevelt agreed, and Michtom produced and began selling "Teddy's Bear" in 1903.

By 1906, the teddy bear was the top-selling toy in America. A year later, nearly one million teddy bears had been sold. Children adored their teddy bears. These fuzzy stuffed animals were made of mohair and came in assorted colors and sizes from three and half to forty inches. Later, Michtom founded Ideal Novelty and Toy, which eventually became the Ideal Toy Company, the largest doll-maker in the world. Teddy bears are still popular with children today.

Questions

Have you ever had a teddy bear? What do you like most about the teddy bear? Why? If you could invent a toy, what would you invent? What would you call it?

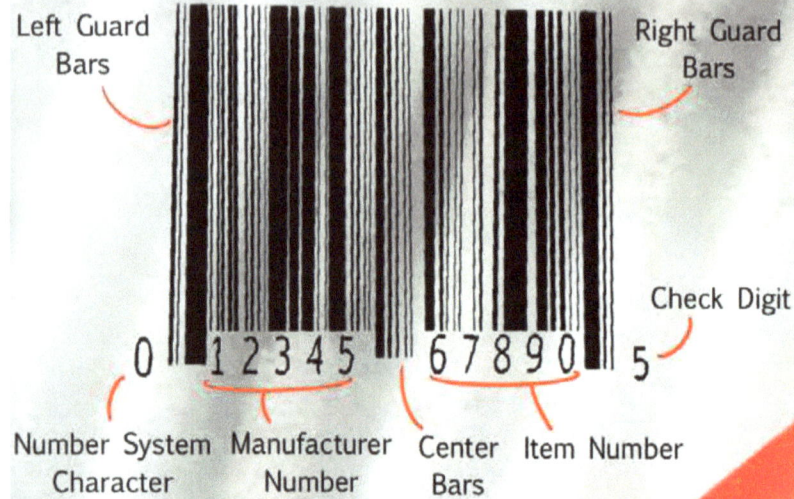

Left Guard Bars

Right Guard Bars

Check Digit

0 12345 67890 5

Number System Character

Manufacturer Number

Center Bars

Item Number

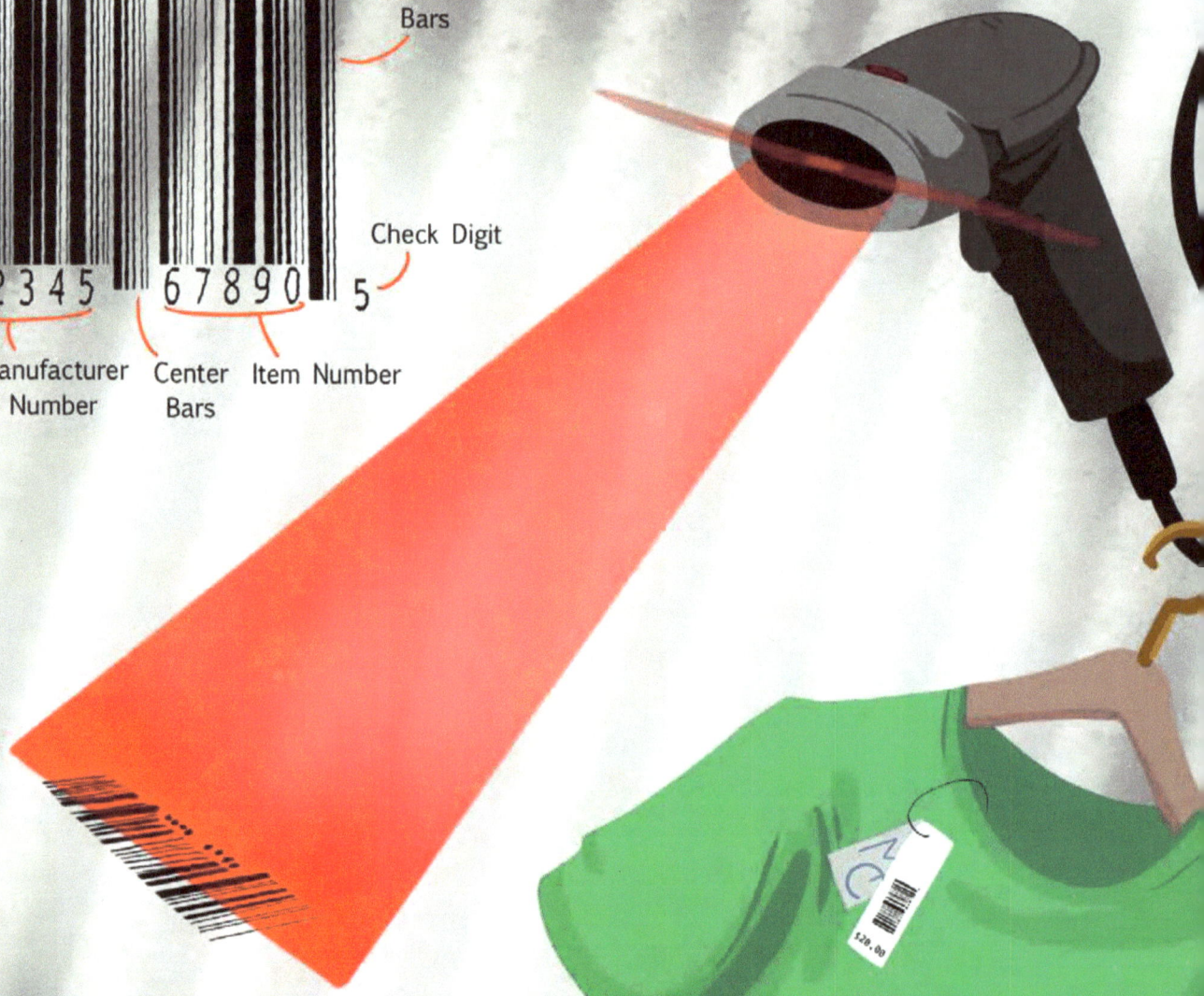

$**3** SAVINGS on your total purchase

Apply coupon at check out. One coupon per purchase. Three dollars off of purhase prices after taxes. Valid on purchases store wide.

Expiration 01/01/10

3 00607 05302 6

Universal Product Code

The first Universal Product Code was developed by George Laurer in 1971. Laurer was an engineer at IBM, and his boss told him to create a circular bulls-eye code that would be compatible with a scanner for checking out groceries. Laurer knew that the circular code would cause problems, so he designed a code with stripes, and it worked.

Laurer's invention became the basis for the Universal Product Code (UPC), and the first UPC or bar code was introduced to grocery stores in 1973. It revolutionized the retail industry. Cashiers no longer had to manually key in the prices of products, because they could be read by the scanner. The bar code also included additional important information such as the manufacturer's identity, product size, and place of origin. Today, the Universal Product Code (UPC) is used on inventory items, railroad cars, library books, pieces of luggage, retail and non-retail items, and many more products.

Questions

Can you imagine what it would be like today if products did not have the Universal Product Code (UPC) on them? How does the Universal Product Code help different industries? Why is the Universal Product Code so important?

Vending Machine

Vending machines have been around for many years. However, Thomas Adams, founder of the Adams Gum Company, is credited with popularizing them in the United States. He patented a machine to dispense gum in 1871. In 1888, Adams selected a train platform in New York City and installed a vending machine filled with his Tutti-Frutti chewing gum. Soon, his machines could be seen on train platforms throughout New York City.

Today vending machines are found in many locations and are used to dispense many different products. Some of the products dispensed by vending machines include water, fruit juices, snacks, fruits, soft drinks, coffee, DVDs and CDs.

Questions

Name some locations that you have seen vending machines. Can you think of any other places where vending machines are needed? Can you think of any other products that can be sold in vending machines? If you could design a new vending machine, what products would be sold in it?

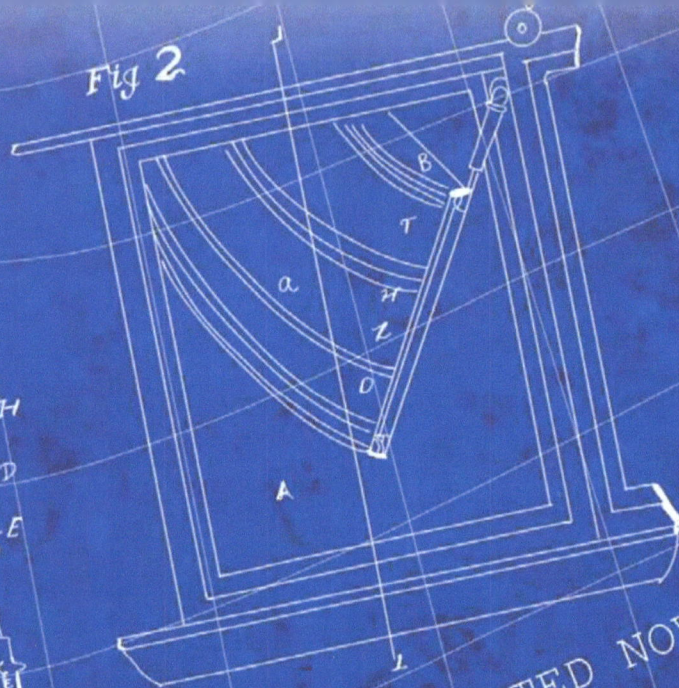

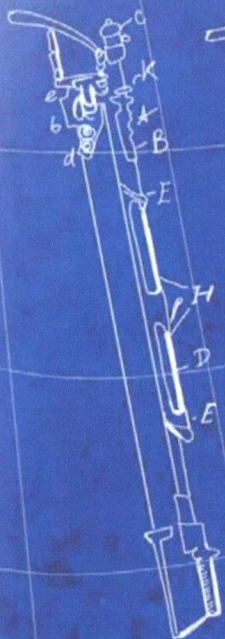

Fig 2

Fig 1.

PATENTED NOV. 10, 1903

Windshield Wipers

While visiting New York in 1902, Mary Anderson noticed that streetcar drivers were having trouble driving in the snow and sleet. They would have to keep the windshield up or make frequent stops to clean the sludge off so that they could see. When Mary returned home, she designed a device with a swinging arm and a rubber blade to clean the windshield. It was controlled by a lever on the inside of the car. She called her new invention a "window cleaning device."

Mary received a patent for her window cleaning device on November 10, 1903, but when she tried to sell it in 1905, no one was interested at first. However, by 1916, windshield wipers were standard equipment on all automobiles.

Questions

How are windshield wipers controlled today? How do they help drivers? Can you imagine what it would be like to drive without windshield wipers? Can you think of anything to add to the modern-day windshield wiper?

Xtracycle

Ross Evans, an engineer, was inspired to design the Xtracycle while he was working on a project called "Bikes not Bombs" in Nicaragua in the 1990s. The Bikes Not Bombs movement began in the early 1980s to provide people in developing countries with a means of transportation and to teach them skills to become bicycle mechanics.

The Xtracycle was designed to carry cargo and helped to provide opportunities for farmers. They could use these cargo-carrying bicycles to transport their products to markets. Another motivation for the invention of the Xtracycle was to get people riding bicycles again. People could exercise while transporting goods in the cargo attachments.

The Xtracycle has an extended frame that allows it to carry a total weight of 400 pounds including the rider. The design of this bicycle has made it possible to carry cargo, exercise, and have an environmentally safe means of transportation.

Questions

Have you ever seen an Xtracycle? Why is it unique? Do you think if more people rode Xtracycles for short commutes it would decrease the use of gasoline and reduce pollution? How could this help our environment?

Yield Sign

Clinton Riggs, an Oklahoma police officer, designed the yield sign while he was a student at Northwestern Traffic Institute in 1939. He designed it to alert drivers to slow down, or sometimes come to a full stop as they approached an intersection. However, the first yield sign was not installed until 1950 at the dangerous intersection of First Street and Columbia Avenue in Tulsa, Oklahoma. In the year after its installation, there were fewer accidents at this intersection. The decrease was so significant that the city of Tulsa placed yield signs in other locations around the city.

The yield sign is a triangular sign that is posted on roads instead of a stop sign. It warns drivers to approach an intersection slowly and proceed with caution. Today, all yield signs are red with a white triangle in the center and red writing. Yield signs are used all over the world.

Questions

How does the yield sign help drivers? Do you think a new type of traffic sign is needed today? Can you design a traffic sign to better help drivers?

Zipper

Whitcomb L. Judson, an American mechanical engineer, invented the "clasp locker" which is similar to our modern-zipper. It is said that Judson's inspiration for the clasp locker was a friend's complaint of a backache from doing up his boots. The button-style boots worn during the 1800s had to be fastened by hand. Judson's invention made it easier for people to fasten their boots. Judson was granted a patent for his clasp locker in 1893. He displayed it at the Chicago World's Fair that same year.

B. F. Goodrich added the fasteners to their rubber galoshes in 1923. Goodrich was the first to call the fastener a zipper. Also during the 1920s, zippers were added to men's pants and later to women's clothing.

Questions

What might have happened if the zipper had not been invented? What kind of shoes have zippers on them today? Are the zippers for fashion or a necessity? Can you think of any other ways zippers can be used?

W.L. JUDSON

Patented APRIL, 23,1905

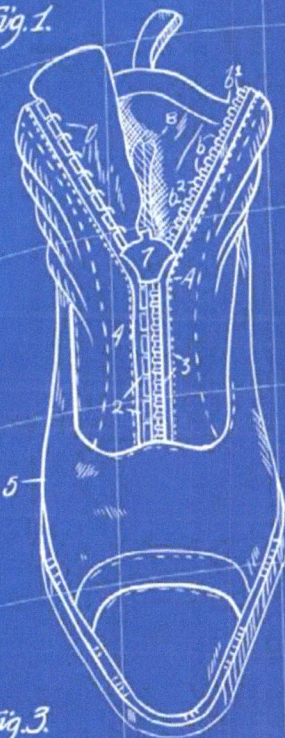

Fig.1.

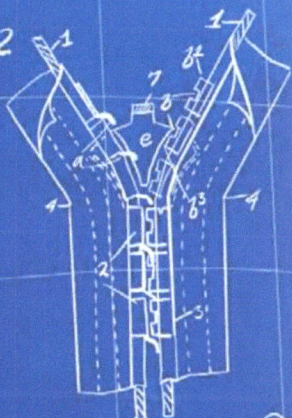

Fig.2

Fig.5.

Fig.6.

e

Fig.3.

Fig.4.

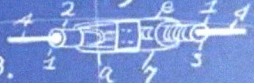

Fig.7.

Fig.8.

Fig.9.

W.L. JUDSON

Patented AUG. 29, 1893

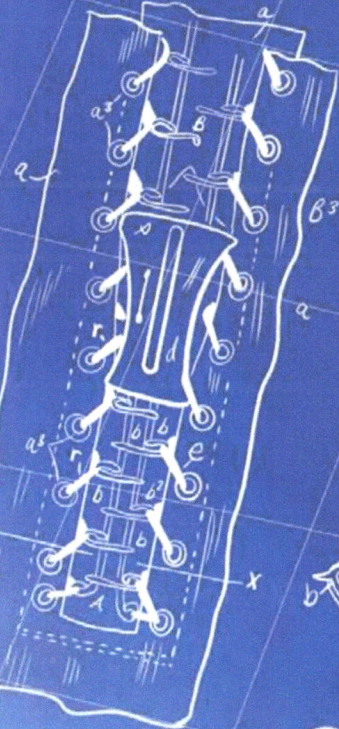

Fig.1.

Fig.3

Fig.4.

Fig.5.

Fig.6.

www.ingramcontent.com/pod-product-compliance
Lightning Source LLC
Chambersburg PA
CBHW050911210326
41597CB00002B/91